Our Animal World

Animal Clothing

by Emily C. Dawson

amicus readers

1

Amicus Readers are published by Amicus
P.O. Box 1329, Mankato, Minnesota 56002

Printed in the United States of America at Corporate Graphics,
North Mankato, Minnesota.

Library of Congress Cataloging-in-Publication Data
Animal clothing / by Emily C. Dawson.
p. cm. – (Amicus readers. Our animal world)
Includes index.
Summary: "Compares different animal coverings such as shells, feathers, fur, and scales. Includes comprehension activity"–Provided by publisher.
ISBN 978-1-60753-006-0 (library binding)
1. Body covering (Anatomy)–Juvenile literature. I. Title.
QL941.D39 2011
591.47–dc22

2010007461

Series Editor Rebecca Glaser
Series Designer Kia Adams
Photo Researcher Heather Dreisbach

Photo Credits

Alexander Kuzovlev/123rf, 4–5, 21 (m); Corbis, 1, 12, 20 (t), 22 (mr); Corbis/Tranz, 14, 20 (b), 22 (bl); Danita Delimont/Alamy, cover; Digital Vision, 16, 21 (t), 22 (br); Gamutstockimagespvtltd/Dreamstime.com, 18, 21 (b); JB Parrett Photography/Getty, 8, 22 (tr); Keith Levit/Shutterstock, 6–7, 20 (m), 22 (tl); Photodisc, 10, 22 (ml)

1224
42010

10 9 8 7 6 5 4 3 2 1

Table of Contents

Animals don't need clothes.
They have many different
coverings that protect them.

Polar bears have thick fur. Each hair is hollow. Their fur traps air and keeps them warm.

skin

Frogs have smooth skin.
They can breathe and drink
water through their skin.

shell

A crab has a hard shell that protects its body. When the crab grows bigger, it sheds its shell and grows a new one.

feathers

Peacocks have long feathers with bright eyespots. Their flashy feathers help them attract a mate.

quills

Porcupines have sharp quills. Porcupines raise their quills when they are attacked.

scales

Snakes have scales that overlap. When they grow, they shed their skin and grow new scales underneath.

People are animals too.
What kind of covering do
you have?

Picture Glossary

feathers
the light, fluffy parts that cover a bird's body

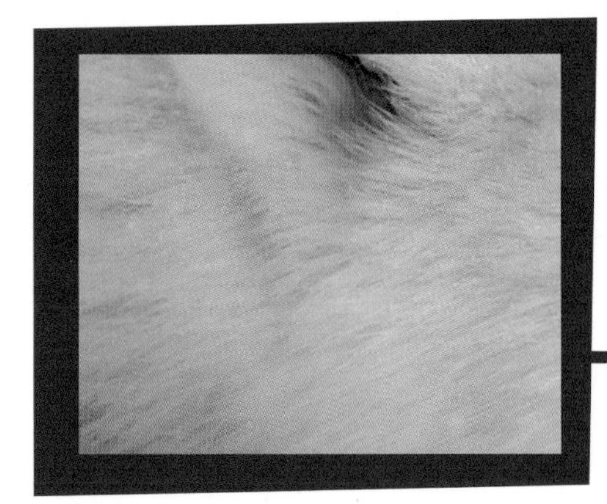

fur
the soft, thick, hairy coat of an animal

quills
long, pointed spines on a porcupine

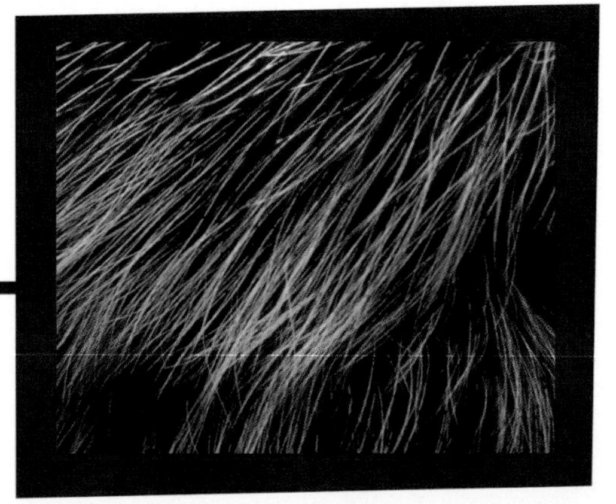

scales
the small pieces of hard skin that cover snakes or other reptiles and also fish

shell
a hard outer covering of an animal such as a crab or snail

skin
the outer covering of an animal

What Do You Remember?

Match each animal to the type of covering it has.

polar bear

frog

shell
fur
feathers
skin
scales
quills

crab

peacock

porcupine

snake

Ideas for Parents and Teachers

Our Animal World, an Amicus Readers Level 1 series, gives children fascinating facts about animals with lots of reading support. In each book, photo labels and a picture glossary reinforce new vocabulary. The activity page reinforces comprehension and critical thinking. Use the ideas below to help children get even more out of their reading experience.

Before Reading

- Read the title and have the students talk about the cover photo. Ask the students if they think animals wear clothes.
- Ask the students what clothes do. What do animals have that do the same thing as clothes?
- Ask the students to turn to the picture glossary. Read and discuss the words.

Read the Book

- "Walk" through the book and look at the photos. Let the students ask questions about the photos.
- Read the book to the students, or have them read independently.
- Show students how to read the labels and refer back to the picture glossary to understand the full meaning.

After Reading

- Have students compare the different coverings. Ask questions, such as *Why does a polar bear not have sharp spines? Why does a snake not have thick fur?*
- Ask the students to think about other animals not in the book. *What types of coverings do they have? How do their coverings help them?*

Index

Web Sites

Animal Games: Animals of the World
http://www.kidscom.com/games/animal/animal.html

Creature Features from National Geographic Kids
http://kids.nationalgeographic.com/Animals/CreatureFeature/

What Kind of Animal Is This? Online Animal Games for Kids
http://www.sheppardsoftware.com/content/animals/kidscorner/kidscorner_games.htm

11841101R00189

Made in the USA
Middletown, DE
15 November 2018